Exotic Animal Farts

Dave Madden

Dedicated to all of "THOSE" crazy office conversations. This book was inspired by one of them. Thanks Fran!

CONTENTS

INTRODUCTION

Hey there, adventure-seeker! Get ready to embark on a wild and wacky journey into the world of "Exotic Animal Farts"! This isn't your average science book; it's a rollercoaster ride through the animal kingdom, exploring one of nature's most hilarious and intriguing phenomena - animal farts!

Did you know that animals fart just like us? But their farts are way more interesting (and sometimes smellier!) than ours. From the roaring farts of lions to the silent, but deadly, releases of snakes, we'll dive into the smelly science behind these toots. And trust us, it's not just about the laughs (although there will be plenty of those!). We're also going to learn some pretty cool science stuff along the way.

In this book, we'll uncover the mysteries of how and why different exotic animals fart. Each chapter will introduce you to a new animal and their unique gassy habits. You'll discover how a zebra's fart can sound like a honking horn and why a kangaroo's farts are good for our planet. It's all about the wild, weird, and wonderfully stinky world of animal farts!

So, buckle up and hold your noses, because we're about to journey through the jungle, dive into the oceans, and leap across the savannahs, all in the name of science! Get ready for some explosive fun - this is "Exotic Animal Farts"!

CHAPTER 1: THE ROARING FARTS OF LIONS

Roar! When you think of lions, you probably imagine their mighty roars. But have you ever wondered about the sounds you *don't* hear? That's right, we're talking about lion farts! Just like their roars, lion farts are a force of nature.

Lions, being big carnivores, eat a lot of meat. This diet means their digestion is pretty heavy-duty. When they digest meat, it creates gas as a by-product - and that gas has to go somewhere! While lion farts might not be as loud as their roars, they're certainly a smelly signal to other animals. In the lion world, a fart can say a lot – from marking territory to signaling that it's time to move on.

So next time you hear a lion roar, just remember: there might be some silent but deadly action happening too!

CHAPTER 2: ELEPHANT GAS EXPLOSIONS

Imagine an animal so big, its farts sound like a trumpet! Welcome to the rumbling world of elephant gas explosions. Elephants are the largest land animals, and their digestive systems are equally massive. When these gentle giants eat, they consume huge amounts of plants, which ferment in their long intestines. This fermentation produces a lot of gas, leading to some seriously loud and smelly farts!

Elephant farts are more than just a cause for giggles; they're a natural by-product of their vegetarian diet. These farts can be so forceful that they've been known to startle birds, making them fly away in surprise. And the sound? It's like a deep, rumbling growl echoing through the jungle. Next time you're near an elephant, listen closely (but maybe don't stand too close behind!).

So, there you have it, the thunderous, trumpet-like toots of elephants, a booming reminder of nature's grandeur (and humor)!

CHAPTER 3: MONKEYS AND THEIR MUSICAL TOOTS

Swinging through the trees, monkeys are not just playful but also quite musical with their toots! Each monkey species has its own unique farting pattern, like a signature tune. The variety in their farts is as diverse as their personalities and habitats.

Monkeys eat lots of fruits, leaves, and sometimes insects, which leads to a symphony of digestive processes. This diet makes their farts a common and often noisy occurrence in the jungle. Some monkey farts are quick and high-pitched, while others are long and rumbling. It's like each troop has its own farting rhythm, adding an amusing soundtrack to the forest!

So, while monkeys chatter and howl in the treetops, their musical toots add an unexpected layer to the jungle's harmony. It's a reminder that in the animal kingdom, even farts can be a form of art!

CHAPTER 4: THE SILENT BUT DEADLY FARTS OF SNAKES

Slithering silently, snakes are masters of stealth, and this includes their farts! Yes, snakes do fart, although it's a rare and often unnoticed event. These silent but deadly releases are as mysterious as the snakes themselves.

Snakes eat less frequently than many other animals, but when they do, they consume their prey whole. This big meal takes a long time to digest, leading to the buildup of gas in their intestines. Unlike the loud farts of larger animals, snake farts are usually quiet, if not entirely silent, and can easily go unnoticed in the wild.

So, the next time you see a snake, remember, it's not just their bite you might not hear coming. Their farts are just as stealthy!

CHAPTER 5: THE THUNDEROUS UNDERWATER FARTS OF WHALES

Dive deep into the ocean, and you might just encounter the thunderous underwater farts of whales. These giants of the sea take farting to a whole new level! Whale farts are not just big; they're colossal, creating bubbles so large they can be seen from the surface.

Whales consume a diet rich in fish and krill, which gets digested in their enormous stomachs, producing a significant amount of gas. When they release this gas, it creates an underwater spectacle. These farts are not only massive but also play a role in the ocean ecosystem, releasing nutrients that support marine life.

So, while whale farts might not be heard above the waves, beneath the ocean's surface, they're like loud, bubbly bursts echoing through the deep blue. It's a reminder of the hidden wonders of marine life!

CHAPTER 6: GIRAFFES AND THEIR HIGH-ALTITUDE RELEASES

Giraffes, with their towering height and long necks, take farting to new heights—literally! These gentle giants have a unique way of releasing gas, making their farts quite the high-altitude affair.

Their diet mainly consists of leaves, especially from acacia trees, which are tough to digest. As these leaves ferment in the giraffe's long digestive tract, they produce gas. Due to their height, when a giraffe farts, it's almost like a gust of wind swooshing down from the treetops!

Although not as loud as some other animals, giraffe farts are a natural part of their digestion. So, next time you're gazing up at these majestic creatures, remember, they're not just eating leaves up there, they're also discreetly passing gas!

CHAPTER 7: ZEBRA STRIPED FARTS

Zebras, with their striking stripes, bring a unique pattern to everything they do, even their farts! These horse-like animals have a playful approach to passing gas. Their farts can sound like a series of honks, almost as if each stripe comes with its own note.

Feeding on a variety of grasses, zebras have a digestive system that ferments these fibrous foods, leading to the production of gas. When a zebra farts, it's often loud and abrupt, echoing across the savannah. In zebra herds, farting can seem almost like a communal activity, with one fart often setting off a chain reaction.

So, in the world of zebras, farts are not just a natural bodily function but also a part of their social interaction. It's a stripe-filled, honking chorus out in the wild!

CHAPTER 8: THE SMELLY LAUGHTER OF HYENAS

Hyenas are known for their distinctive laughter, but there's another, smellier side to their communication – their farts! These cunning creatures have a digestive system well-adapted to processing bones and tough meat, which leads to particularly smelly farts.

Interestingly, hyenas' farts are often released during moments of excitement or stress, such as when they're competing for food or feeling playful. This means that a cackle of laughing hyenas might be accompanied by a chorus of farts!

Their farts serve as a smelly signal in the hyena community, communicating everything from territorial claims to social status. So, in the hyena's world, a fart isn't just a funny sound, it's an important part of their wild conversation!

CHAPTER 9: KANGAROO HOPS AND POPS

In the Australian outback, kangaroos are famous for their impressive hops, but there's another, less known aspect to their leaps: their farts! As kangaroos hop around, their movement aids in releasing gas built up during digestion.

Kangaroos have a unique diet of grasses and shrubs, which ferment in their specialized stomachs, leading to the production of methane-free gas. This means kangaroo farts are not only frequent but also environmentally friendly!

Each hop can be accompanied by a small pop, making their movement not just a visual spectacle but a humorous symphony as well. In the world of kangaroos, farts are a natural, eco-friendly part of their hopping lifestyle.

CONCLUSION

And that's the end of our fragrant journey through the animal kingdom! From the roaring farts of lions to the eco-friendly pops of kangaroos, we've explored how each animal's unique diet and lifestyle contribute to their distinct farting habits. We've laughed, wrinkled our noses, and most importantly, learned that farting is a natural and necessary part of animal life.

These gassy tales remind us of the diverse and wonderful ways nature works. So, the next time you hear or smell a fart, remember the amazing animals we've met and their special toots. Farting, in all its forms, is just another remarkable aspect of our incredible natural world. Keep exploring, keep learning, and keep giggling!

GLOSSARY

Carnivore: An animal that eats meat.

Digestion: The process by which an animal's body breaks down food into nutrients.

Ecosystem: A community of living organisms and their environment, working together as a system.

Fermentation: A chemical process where bacteria break down substances such as sugar without the use of oxygen, often producing gas.

Fiber: Parts of plants that cannot be digested, which help to keep the digestive system healthy.

Gas: An air-like substance which can expand freely and is used in the book to refer to farts.

Giraffe: The tallest living land animal, known for its long neck and legs.

Hyena: A carnivorous animal known for its strong jaws, distinct calls, and scavenging habits.

Intestines: Long, tube-like organs in the body where most of the digestion and absorption of food takes place.

Kangaroo: A marsupial from Australia, known for its strong hind legs and hopping movement.

Krill: Small crustaceans found in the ocean, which are a significant part of the diet for many marine animals, including whales.

Lion: A large wild cat known as the "king of the jungle," famous for its roar.

Methane: A colorless, odorless flammable gas that is a product of digestion in many animals.

Nutrients: Substances that provide nourishment essential for growth and the maintenance of life.

Ocean: The vast body of salt water that covers almost three-quarters of the earth's surface.

Savannah: A grassy plain in tropical and subtropical regions, with few trees.

Scavenger: An animal that feeds on carrion, dead plant material, or refuse.

Snake: A long, legless, carnivorous reptile.

Species: A group of living organisms consisting of similar individuals capable of exchanging genes or interbreeding.

Territory: An area defended by an animal or group of animals against others of the same species.

Vegetarian: An animal that eats mostly plants.

Whale: A large marine mammal that breathes air through a blowhole on the top of its head.

Zebra: An African wild horse with black-and-white stripes and an erect mane.

DID YOU KNOW?

Did You Know? #1:
Lion's Roar and More!
A lion's roar can be heard up to 5 miles away, but their farts are much more silent and secretive!

Did You Know? #2:
Elephant Fart Facts!
An elephant's digestive system can produce enough gas to fill a large trash bag every single day!

Did You Know? #3:
Monkey Business!
Some monkeys have such strong farts that they can be used to scare away predators or unwanted company!

Did You Know? #4:
Snakey Silent Farts!
Snakes fart so infrequently that when they do, it's a rare occurrence for wildlife biologists to record!

Did You Know? #5:
Whale Bubble Nets!
Whales sometimes use farts to create bubble nets, which help them herd fish for easier feeding!

Did You Know? #6:
Giraffe Winds!
A giraffe's fart has to travel a long way down before it can be released, due to their tall stature!

Did You Know? #7:
Zebra Stripes and Sounds!
Zebras have unique farting patterns, just like their stripes – no two

zebra farts are exactly the same!

Did You Know? #8:

Hyena's Laughing Gas!

Hyenas' giggles and guffaws are often accompanied by farts, making their interactions even more comical!

Did You Know? #9:

Kangaroo Methane!

Kangaroos don't produce methane when they fart, making them eco-friendly animals, unlike cows!